Let's Show Your
Mule

A Mule Photo Album
1993

By
Robert A. Mischka

Heart Prairie Press
Whitewater, Wisconain

Published by:
Heart Prairie Press
P. O. Box 332
Whitewater, WI 53190
(414) 473 5595

Printed in Hong Kong

Publisher's Cataloging in Publication Data
Mischka, Robert A. 1934 -
Let's Show your mule: a mule photo album, 1994 / by Robert A. Mischka
p. cm.
ISBN 1-882199-02-2

1. Mules--United States--Pictorial works. 2. Mules--United States--Exhibitions--Pictorial works. I. Title.
SF632.M57 1994 636.18

Front Cover Photo: The color guard during the opening ceremonies at the Bishop Mule Day grandstand show.

Back Cover Photo: Don Bales driving his six of Belgian mules in the Great Milwaukee Circus Parade

Introduction

The enthusiasm with which mule folks greeted our Mule Calendar in 1991 made us aware of the need for more mule information, especially information which avoids the common stereotypes which poke fun at mules. While I have worked with saddle horses and draft horses for most of my adult life, I am not a mule man. But I enjoy taking pictures, and have become fascinated with the beauty and character of mules. This book is an attempt, along with our calendar, to show the usefulness, personality, versatility, and beauty of today's mules.

Today's mule is quite different from those bred in the early part of this century. Better jacks are being mated with outstanding mares of all breeds to come up with mules for a variety of specific purposes. These new mules are making their mark in endurance riding, dressage exhibitions, carriage driving shows, racing, trail riding, and many other endeavours formerly the exclusive province of the horse.

The very diversity of the mule makes it difficult to write a book which appeals to all mule people. In trying to include all parts of the mule community (draft, riding, racing, and packing) we risk pleasing no one. This is a risk we take in our Mule Calendar, and which I have taken in this book.

Mule people tend to be friendly, fun-loving people. For years they have been the butt of jokes from their horse friends and the general public. Their own confidence in their choice and their strong feelings toward their animals makes them ignore these taunts from people who just plain "don't know any better".

But constantly defending your choice of animal with the uninformed can become tiring at times. So mule folks take their animals to mule shows where they can just enjoy themselves with others who share their enthusiasm for mules, without any apologies. I attended three of the larger mule shows in the United States to take pictures for this book. The United States is so big that most Tennessee folks can't make it to the California show, and vise versa. So for those of you who weren't able to go I made a point of going for you. And for those who did go, I hope you will enjoy remembering the experience through these pages.

A mule show is different from most horse shows. Having fun and enjoying yourself is the primary, and in some cases the only, objective. With the diversity of the mule world displayed in a broad variety of classes there is something for everyone. You aren't into dressage? Then go watch the mule pull. Pleasure classes bore you? Then watch the chariot races. But, for heaven's sake, don't miss the packer free-for-all.

But along with this "fun and games" attitude there is a growing attempt on the part of the saddle mule community to bring mule shows up to the quality and standards set by the major horse breeds at their breed shows. Saddle mules are growing in acceptance and popularity. They are the fastest growing segment of the mule industry. It is through saddle mule shows that this growth will continue and intensify. There are many in the saddle mule community who are working hard to upgrade the saddle mule classes at mule shows, and their efforts can be seen at two of the four events covered in this book.

I hope you will enjoy this celebration of the mule as we visit some of the 1993 Mule Shows.

Bob Mischka

Acknowledgements

Many people have helped me with this book. Being a relative newcomer to the world of mules, I needed help in identifying people and in locating them to get and verify information for the photo captions.

I owe special thanks to Loyd Hawley, John Robert Skillington, Don Jacklin, Judy Rose, Marilyn Briggs, Jennifer Jones, and Cathy Neal. Without their help the book would have been more incomplete, and less interesting. Any identification mistakes which the reader may find are mine, and are unintentional.

Table of Contents

Joe Frank Skillington *led off the 1993 Columbia Mule Day parade with a team of sorrel mules pulling a restored John Deere wagon bearing the flags of the United States, the State of Tennessee, and Maury County Tennessee. With Joe is his sister, Faye Bohannon of Houston, Texas. Joe, now age 63, drove a 20-mule team in the 1950 Mule Day parade, a feat which had not been equaled until 1993 when a group put together a 22-mule team. For information and pictures of the 1993 20-mule team see page 18.*

Mule Day

Columbia, Tennessee

About 50 miles southwest of Nashville you will find the little town of Columbia, Tennessee, now best known as the home of the General Motors Saturn plant. But each year, on an early April weekend, Columbia forgets all about Saturn, and just about everything else, and proclaims itself the Mule Capital of the World. They call the event Mule Day, and it consists of a big parade on Saturday surrounded by various smaller activities starting Thursday night and ending Sunday evening. The main emphasis during this event is on draft mules.

Today's Mule Day origins go back in the mid-1800s when local farmers would gather to buy and sell mules on the first Monday of April, called Breeder's Day. This evolved into a Livestock Show Day in the late 1800s, and a parade was added. In 1940, the celebration was called "Columbia's Big First Monday Breeder's Day and Mule Day". The event was dropped in 1950 as being out-moded, only to be started up again in 1974 under the sponsorship of the local saddle club. It now seems to be well established, with large crowds watching the Mule Day parade each year.

Western Tennessee can be a beautiful place in early April. And on Mule Day the place to be is Columbia. It's a time to renew old acquaintances, to see what the other fellow has been up to and has in the way of mules, and to shake off the inactivity of the winter months. It's a celebration of Spring and a celebration of the mule, especially the draft mule.

A team of mules owned by **Bernis and Janice White** *of Santa Fe, Tennessee, leaves the Maury County Park, headquarters for the Columbia Mule Day celebration, on their way to the Saturday parade.*

DRAFT MULE SHOW

The draft mule show is divided into two parts: the hitch show on Friday afternoon; and the halter show on Saturday, after the parade. They don't call it a *draft mule* show, but that's what it is. There is a class for other harnessed mules on Friday, and a class for saddle mules on Saturday, but they are more in the way of token classes with only three entries each day.

W. F. Reed *of Franklin, Tennessee, driving his mule in the single mule log skidding class.*

The hitch show starts off with two log skidding classes, one for a single mule and one for a team. The mules in these classes are driven with a single line to the left side of the bridle, or with no line at all. The mule(s) are hooked to an empty pulling sled (rather than a log), and the teamster talks his mule(s) through a serpentine pattern around some cement blocks. Following these two classes there are the usual single, team, unicorn, four and six horse hitch classes, a farm team class, and a driving skill class. The judge asked each exhibitor for an individual workout in each class, which turned each class into a driving skill class, and extended the show into the evening hours. The class with the largest number of exhibitors was the Show Team Class with eleven Belgian mule teams and one Percheron mule team.

The halter show on Saturday is serious business, with a total of twenty-seven individual classes. The show starts with three Jack classes, followed by a series of mare mule and horse mule classes for both individuals and pairs, starting with yearlings and going to four and over. The mare mule classes were well filled, with eleven yearling mare mules and twenty-one 4-yr-old mare mules. The judge, John Chitman, had a long afternoon's work — which, in fact, went well into the evening both days.

Thomas Summers *of Morrison, Tennessee, talks his 5-yr-old mule ,* ***Kate****, through the single mule log skidding class, without using a line.*

Wayne Clark *of Erin, Tennessee, also used no lines — he just talked his mules to a first place in the team log skidding class. His mules are* ***Jim*** *and* ***Dan****, 7- and 8-yr-old full brothers.*

Earl Wayne Tomlinson *of Bon Aqua, Tennessee, used a single rope which branched to the near side of each mule and a jockey stick to drive the team owned by Clifford Davidson in the log skidding class.*

Jerry Brown, *the hitch-wagon builder from Dexter, Missouri, with his team in the show team class. The mules are 4-yr-old half sisters,* **Kit** *and* **Pat***.*

McCrary Farms of Lebanon, Tennessee, had the only Percheron mules in the hitch show, a pair of 17 hand 4-yr-olds named **Jake** *and* **Kit***.* **Eddie McCrary** *is driving, and Tommy Jones is riding.*

__C. C. Sewell__ of Cookeville, Tennessee with his team of molly mules, __Lize__ and __May__. This team was later judged as the champion pair of mules in the halter show. Riding with C.C. is his daughter, Betsy Hanks.

Left: __Buddy Black__ of Woodbury, Tennessee, drives his 3-yr-old mules, __Matt__ and __Mag__, hooked to a lovely hitch wagon built by his passenger, Bobby White of Mobile, Alabama.

The entry of __C. B. White__, Bowling Green, Kentucky, made a pretty picture as they did their figure eight in the individual workout. The mules are 4-yr-olds named __Mindy__ and __Mandy__. C. B. (Charles) is doing the driving.

Thomas Summers *won the show team class with his 5-yr-old mules,* ***Kate*** *and* ***Alice****. His daughter, Lori, held onto her hat in the cold wind. Tom did very well in the hitch show, winning four firsts and two seconds.*

W. F. Reed *going through his swing pattern in the individual workout, with the mules calmly waiting for him to ask them to swing back.*

Greg Tabor *of Scottsville, Kentucky, showed this molly mule called* **Jane** *to a first place in the yearling mare class. Since it was early April the mules are still wearing their winter coats.*

This business of getting yearling mules to look alert, with their ears forward, can be hard work. The judge, **John Roy Chipman** *of Perry, Missouri, stopped to take another look at both the mules and their handlers,* **L.C. and Greg Tabor**.

Three mules owned by **Fred Flippen** *of Lebanon, Tennessee.*

This pair of 4-yr-old mare mules, ***Kate*** *and* ***Iola****, took third in their class for* **L. C. and Greg Tabor**.

This is ***Lize****, the first place mare mule aged 4 and older, owned by C.C. Sewell and shown here by his grandson,* **Bob Kaiser**. *This mare went on to be named Grand Champion Mule of the Show.*

THE MULE DAY PARADE

The main event during the Columbia Mule Day weekend is the Saturday Mule Day Parade. Thousands of people line the parade route to watch 250 separate parade units in the two hour-long parade. Each entry in the parade must be pre-registered by the Mule Day office, and their position in the assembly area is assigned in advance.

The parade begins at 11:00 am, but all entries must be in the assembly area an hour earlier so they can be judged. Cash awards are given in eight different catagories. The fairgrounds where the mules are stabled, and where most of the other Mule Day activities are held, is about four miles from the spot where the parade starts, and another four miles from where it ends.

Mule Day, 1993, was a lovely, warm, sunny day — a perfect day to watch a parade and get a little sunburn. I enjoyed it thoroughly.

***Kit** and **Dan,** 3 and 4-yr-old full brothers owned and driven by **Larry Dyer** of Petersburg, Tennessee.*

John Robert Skillington *of Santa Fe, Tennessee, with his team. John Robert had one of the few wooden wheeled hitch wagons in the parade. Riding with John Robert is Glenda Bowman and her 19 month granddaughter, Rachel Serkownek.*

Jerry K. Brown *from Dexter, Missouri, with his four-up of fancy mules. The wheelers are* ***Kit*** *and* ***Pat****, also pictured on page 10. The leaders are* ***Katie*** *and* ***Kate****, 6-yr-old half sisters. Jerry takes Katie and Kate on wagon train rides totalling 500 miles each year. Jerry's passenger is Dale Futrell of Hazel, Kentucky.*

Local and state politicians rode many of the wagons and floats. The Commissioner of Agriculture, L.H. Ivy, rode the seat with **Thomas Summers** *of Morrison, Tennessee.*

The wheelers on **Jack Strode***'s four-up were a pair of coming 2-yr-olds — they will be very big mules when grown. Jack, from McMinnville, Tennessee, is an outstanding teamster who frequently shows his mules at the big Detroit International Draft Horse (and Mule) Show — and often wins every hitch class.*

One of the highlights of the 1993 Mule Day parade was this ***20 mule team*** *put together by a group of East Tennessee farmers known as the Mule Skinners. The top picture shows the 20 mule team leaving the fairgrounds on the way to the parade assembly area. The bottom picture shows how the hitch looked during the parade, with riders on each outside lead mule and the addition of two more mules in the lead, making it a 22 mule team. The basic 20 mule configuration for this hitch was five spans of four abreast which, of course, was different from the old 1880s historic Borax 20 mule teams which consisted of ten pairs.* **Wooley Davis** *of Bulls Gap, Tennessee, the man riding the mule in the center of the bottom picture, was one of the main organizers of this team.*

The **Bernis and Janice White** *team from Santa Fe, Tennessee were sporting straw hats for the parade. This is* **Jack** *and* **Dan,** *a pair of 9-yr-olds who have been in the parade every year since they were two.*

Dee C. Neeley *of Columbia, Tennessee, drove a pair of black 2-yr-old mules hooked to grain box mounted on a rubber-tired running gear. Here they pass in front of the Church-Sweeney House, built in 1872 and now on the National Register of Historic Buildings.*

This six-mule hitch of blacks was the entry of **Hugh and Laura Moffett** *of Fayetteville, Tennessee. Hugh is driving and his brother, Bill Moffett, is riding.*

Blythe Woodside *of Franklin, Tennessee, leaving the fairgrounds for the parade with his 3-yr-old mare mules,* ***Pat*** *and* ***Jane****. Riding with Blythe is his grandson, Travis Woodside.*

The **W. F. Reed** *team made a pretty picture as they left the fairgrounds, right in step.*

Dick Sparrow *brought the* ***Coors six horse hitch*** *of Belgians to the Columbia Mule Day celebration for the 8th straight year. Dick drove the hitch in the parade and gave an exhibition drive on both Saturday and Sunday afternoons. At the close of each exhibition he gallops his team around the track, much to the delight of the crowd.*

Jerry Cunningham *of Maryville, Tennessee, had an eight horse hitch of Belgians in the parade. Riding on the seat with Jerry is Dean Kleckner, President of the American Farm Bureau.*

A team of black mules owned by **Milton Hatton**.

There were a few riding mules in the parade, but not many. **Lisa Holt** *won the Best Costumed Entry Award with her indian maiden outfit on an Appaloosa mule.*

Motorized units are allowed in the parade. This colorfully dressed group of Cloggers entertained the spectators from the back of a moving truck.

Steve Cooper of the Country Western band called Canyon, *was Grand Marshall of the 1993 Parade. He rode in a wagon pulled by the team owned and driven by* **Buddy Black** *of Woodbury, Tennessee.*

Cale Reed, *age 6, was surely the youngest teamster in the parade. Cale, here with his parents,* **Charlie and Jackie Reed** *of College Grove, Tennessee, did a very professional job of driving his father's team.*

Worth Whisnant *made the long trip from Hickory, North Carolina, to drive his young horse mule flanked by two 2-yr-old Brahma steers in the parade.*

The **Pugh** *family from Crab Orchard, Tennessee, drove a more conventional 3-abreast entry — three mules hooked to a massive antique farm wagon.*

MULE PULL

The mule pull is held on Saturday afternoon. There are six separate weight classifications, starting with pony mules at a combined weight of 800# or less and ending with large draft mules with a combined weight of 2,200 and over.

The mules are hooked to sled loaded with cement blocks, with blocks added until there is a winner. The distance the larger mules are required to pull in order to advance to a heavier weight is 10 feet. About 50 mule teams compete at the pull. The grandstand was full all afternoon, with the pull lasting into the early evening.

Perry Barkley *came up from Ripley, Mississippi with his heavyweight team of black mules.*

This is **Lonnie Forrest** *of White Bluff, Tennessee, with his pulling team.*

WESTERN MULE SHOW

On Sunday afternoon they have what is called a Western (Saddle) Mule Show. There is a coon jump class, a western pleasure class, two barrel racing classes, and a bunch of fun-and-games classes. For that matter, the coon jump and barrel racing classes were not conducted seriously — they turned into fun-and-games classes also.

Columbia Mule Day is primarily a draft mule event. There is a relatively small turnout of participants and spectators for this saddle mule show. This show does little to showcase the modern, well-trained, versatile mule — which must be why many well-known saddle mule owners do not attend. This show badly needs the care and attention of someone who is both knowledgeable about mules and committed to presenting them in a more favorable light.

Needless to say, I was disappointed in the Western Mule Show.

The Western Mule Show opens with a class called Grand Entry. It is basically a beauty contest, with the winner being the most appealing or attractive entry. **Adam Vale** *of Jefferson City, Tennessee, riding* **Oscar**, *was an easy winner. Adam is 4 years old, and Oscar, owned by John Steward, is 17 years old.*

Terry Hodges *rode the barrel races bareback.*

There were 14 entries in the Western Pleasure Class, but only a handfull were dressed properly and could put their mule into the proper gait. **Monica Byrd** *of Mt. Pleasant, North Carolina, was one of that handfull.*

Abraham *had a bit of trouble with his donkey,* **Amos,** *in the coon jump class. He was not the only one with trouble, as it took an hour and a half to complete this class. A limit on the time allowed each exhibitor would have helped speed things along a bit.*

Tarbaby, *on the other hand, fairly flew at the jump. Tarbaby is owned, and being handled here, by* **Koy Flowers** *of Spencer, Tennessee.*

The grandstand at the Fairgrounds where the Bishop Mule Days events are held, with the snow-capped Sierra Mountains in the distance.

The commemorative belt buckle designed by **Bonnie Shields** *for the 1993 Bishop Mule Days.*

There is a large number of exceptionally fine booths and commercial displays at Bishop Mule Days. If you need something for your mule you can get it here. This is the **Butte Creek Outfitters** *display.*

Mule Days

Bishop, California

During Memorial Day weekend the "Mule Capital of the World" moves west to the little town of Bishop, California, for their Mule Days celebration.

California is a large state, with many different climates and landscapes. The Sierra Mountains run lengthwise up the length of the state, and Bishop lies at the eastern foothills of the Sierras at an elevation of 4,000 feet, making it seem more a part of Nevada than of California. The town has a population of 3,500, and is the jumping-off point for skiing, fishing, hunting, back-packing, camping and other recreational activities in the nearby mountains. But on Memorial Day weekend, for the past 24 years, Bishop is over-run with mules and mule lovers. Local authorities estimate that in 1993, 40,000 out-of-towners visited Bishop for Mule Days.

We enjoyed cool nights and comfortable sunny days during the 1993 Mule Days, and thought that maybe they were always blessed with this type of weather. But no, we were told that Mule Days can be terribly cold, with snowstorms in the nearby mountains, or as hot as the nearby deserts, with cottonwood seeds six inches deep on the ground. I guess we were lucky.

The primary emphasis at Bishop is on riding and pack mules, with only a sprinkling of draft mules. The show is a wonderful showcase of the versitality of the mule, and of the many types and kinds of mules there can be. It is a "really big show", with a total of 136 classes plus a number of races and parade.

The champions at Bishop are called "World" Champions. At least this is what they are called by the show organizers — and by the winners. While this is a bit presumptuous, until the two mule organizations can get together or until a better show comes along I guess it might be a pretty fair title. They do not crown a World Champion Draft Mule at Bishop, as there are few draft mules there.

The main ring during one of the nightly opening ceremonies. The weekend evening shows are really extravaganzas designed to entertain the large audiences. The pack team in the foreground belongs to **Lee and Jennifer Roeser**, *operators of the McGee Creek Pack Station.*

ENGLISH MULE SHOW

Although the Bishop show isn't divided into catagories such as "English Mule Show", for the purposes of this story we'll group the classes into general catagories like this. The classes pictured in this section are actually spread out over the four-day show.

The English Show begins with riding dressage classes, followed by English Pleasure, Working Hunter, Jumping, and Side-Saddle classes. The mules in these classes are elegant, refined and athletic. The exhibitors are appropriately dressed and excellent riders. Everything is first class — a delight to watch.

Three shots of **Pat Mitchell** *of Sun Valley, California, and her mule,* ***Kitty's Maybelline****. This mule was named 1993 World Champion English Performance Mule after placing well in English Pleasure, English Working Hunter, English Hunter Hack, English Jumping, and Dressage, Level 1.*

Lou Moore *from Templeton, California, takes a jump with* ***Empress Josephine****, 1993 Reserve World Champion English Performance Mule. In 1992 Empress Josephine was overall high point mule at Bishop, giving her the crown of Grand World Champion. Josephine is a 9-yr-old mare owned by Lou out of an Appaloosa mare and by the jack Big John. Josephine took first in the Jumping Class at Bishop in 1993.*

Lou Moore *rode* ***Action Annamule*** *in the English Working Hunter Green Class.*

Tim Phillips *takes a jump with* ***Domino****, owned by Warren Sohrakoff. Domino won the Western Pleasure and Western Riding classes, and went on to be named World Champion Western Performance Mule — a very versatile mule.*

Crystal Elzer *of Scotts Valley, California took first in the Side Saddle class with her mule,* ***Teaspoon****, an 8-yr-old, 14 hand mare by a standard jack out of a quarter horse (cutting) mare. This was Teaspoon's first time in side-saddle — she normally is shown in Western Classes. She placed second in Western Riding at this show — out of 23 entries. In both 1991 and 1992 Teaspoon was the American Mule Association "All Around High Point Mule". Crystal is co-founder of the California Side-Saddle Organization.*

Shell Reid *rides her mule,* ***Ruth****, in the Dressage workout. Ruth is a 6-yr-old mare out of a Texas Quarter horse dam. This pair won the Green English Championship at Bishop in 1992. Shell and her husband, Chuck, raise and train mules at their Burnt Leather Ranch in McLeod, Montana.*

Bobbe Johnson *with* ***Rose Marie*** *in one of the Dressage classes. Rose Marie is a 15 hand mare by Bow Valley Jeremiah out of a Quarter mare, Partee Nita. Bobbe rides Marie in both English and Western Classes. Most recently Marie is being taught to play polo by Sue Sally Hale of the Moorpark Polo Club.*

This is **Meredith Hodges** *jumping* ***Lucky Three Mae Bea C.T.*** *, an 11-yr-old mare by a standard jack called Colorado Poncho, and out of the solid-colored Appaloosa mare Sam's Nickey. In the side saddle class Meredith is riding* ***Lucky Three Ciji****, a 10-yr-old half sister to C.T., being by Little Jack Horner and out of Sam's Nickey. The winnings of these two mares are far too numerous to mention here. Both mares were bred, raised, and trained by Meredith at her Lucky Three Ranch in Loveland, Colorado, where she specializes in dressage and combined driving training. Meredith is an expert mule trainer, and the author of the new book* ***Training Mules and Donkeys.***

Half Assed Brother Charlie *makes it look easy as he clears this hurdle with lots to spare. Charlie, a 15-yr-old out of an Appaloosa mare, is owned by* **Bob and Joanne Brand** *of Las Vegas, Nevada, and is being ridden here by Candee Coffee. Bob bought Charlie at Bishop Mule Days twelve years ago, and for the past four years has been showing him in both English and Western classes.*

Christine Allen *won the Youth Western/English 13 and Under Equitation Class riding this 17.2 hand mule called* ***Actions Whattamule****, owned and trained by* **Kelly Baker** *of Chino, California.*

Right: **Pat Mitchell** *is riding* ***Tea Party*** *for* **Jonnie Greslie** *to a 2nd place in one of the four different dressage classes. Tea Party is a 12-yr-old mare by Redman out of a Morgan mare. She is no stranger to the show circuit, having won World Champion English Performance Mule in 1989 and Reserve World Champion English Performance Mule in 1991. Jonnie Greslie showed her to Overall Champion on the West Coast Inter-Collegiate Circuit in both 1989 and 1990, competing against horses.*

MULE SHOEING COMPETITION

Daytime Bishop Mule Day events take place at five different sites on the Fairgrounds, with several different events going on at the same time. For example, on Sunday morning, spectators have a choice of watching Cow Cutting, Mule Pulling, the Trail Class Finals, the Saddle Mule Halter Classes, or a Mule Shoeing Contest. Some hard choices have to be made.

The Mule Shoeing Classes consist of three classes: a Quality Shoeing Contest, a Forging Contest, and an Eagle Eye Class. The shoer with the best record in these three classes is crowned the World Champion Mule Shoer.

Tony Baratti, *the 1992 World Champion Mule Shoer, cut a fancy figure with his derby, fringed chaps, and tennis shoes. The message on his shirt says "The more explicitly you instruct the farrier the more likely he'll do what he damn pleases". Tony is from Dayton, Nevada.*

COON JUMPING DEMONSTRATION

They didn't really have a class for coon jumping, but two of the exhibitors put on a coon jumping demonstration for the benefit of the crowd.

Jerry Villines *and* ***Booger's Chiquita*** *won the Coon Jumping demonstration. They also won Reserve World Champion Gymkhana Mule after placing second in barrel racing and third in the Keyhole race. At the right we see them perform in the barrel race. Jerry is from Coffeyville, Kansas.*

MULE RACES

There are a total of 15 (riding) mule races at Bishop Mule Days — some on each day of the show. Most of the races are professional races, with an official starting gate and professional jockeys. A few are what they call "Western Races", where the riders are amateurs and the starting gate is not used. The lengths of the races range from 200 to 400 yards. They are over in a flash.

Most of the races were quite close, as this one was. They are over so quickly that the best way to really see them is in a photo. This is one of the professional 200 yard races, with ***Jeff McGee*** *in the center of the picture (white blinkers) on his way to a win. The jockey riding Jeff is Luke Wethey. Jeff McGee, owned by* ***Don Jacklin*** *of Post Falls, Idaho, went on to be named World Champion Running Mule. The mules in this photo,.from left to right, are* ***Loretta Lynn, B. D. Rocket, Jeff McGee, Lucky Penny, and Cajun Queen****.*

Scott Luce rides **My Molly D**, *owned by McMunnia, to a win in one of the Maiden races. Coming in second is* **Avery's Rebel**, *ridden by Bailey and owned by* **Jim Farley**.

Okie Bailey, *with the red blinkers, is leading the pack in this race. Okie is owned and ridden by* **Don Myers**.

DRIVING CLASSES

Bishop has several types of driving classes, including pleasure driving, obstacle driving, chariot racing, donkey driving, and 2, 4, and 6 mule classes. The driving classes are all held on Friday.

These classes feature a wide variety of mules, from fancy light driving mules to heavy draft mules.

*Freddie Valentino of Lancaster, California, won World Champion Driver honors. Above he is driving **Emmy and Kate** in the 2 Mule Hitch Class. At the right he is getting set to enter the Americana Class, a class for wagons with original equipment and appropriate attire, with the same pair of mules.*

Richard Pauley *of Saugus, California, driving his team of black mules,* ***Jack*** *and* ***John****.*

Left: **Dawn Marie** *from Santa Rosa, California, drives her mule,* ***Apples, Peaches and Cream****, in the Gambler's Choice Obstacle Class. Dawn took first in the Pleasure Driving Class with this mule.*

Calico Gal *is completely airborn as she streaks for another obstacle in the Gambler's Choice Obstacle Class. Calico Gal is a 12-yr-old mare by a spotted jack and out of a quarter horse mare. She is owned and being driven by* **Gary Bloxham** *of Salinas, California. They placed 2nd in this class.*

Ruth Jacobsen *from Cayucos, California, took third in the Pleasure Driving Class with her handsome mule,* ***Prince Albert****, a 13-yr-old by the spanish jack Stay Ready Jim, and out of a quarterhorse mare. Ruth and Prince Albert taught each other to ride and drive, with no professional help, six years ago, when they were ages 75 and 7, respectively. They compete in halter, English, Western, driving, and packing classes in California and Utah shows.*

Jon Martin *drove* ***Maude and Jack*** *to a first place in the Chariot Racing Class. The Martin Brothers, from Nampa, Idaho, won 1993 World Champion Teamster honors.*

*This is **Rooster**, four times World Champion Jack at Bishop (1989-1992), being driven by **Charles Moore** in the Donkey Driving Class. Rooster is a 7-year old Mammoth Jack owned by the HME Stock Ranch of San Diego, California.*

***George Lane** drove his pair of driving mules, **Bonnie and Bar**, in the Team Class. The Lane Ranch is located in Lancaster, California. George also drove a four hooked to a stagecoach in the Four Mule Class — we picture the stagecoach on page 67 in the parade section of the Bishop story.*

THE TWENTY MULE TEAM

In 1990 the Bishop Mule Days brought back the 20 mule team, famous for hauling borax out of the desert of Death Valley during the five year period 1883-1889. These teams pulled three wagons, two loaded with 10 tons of borax each and one full of water.

The 20 mules are hitched to an 80 foot chain which is fastened to the lead wagon. The driver uses a single "jerk" line to the left-hand mules, plus the vocal commands of "gee" (left) and "haw" (right), to direct the team.

The driver is Bobby Tanner, the wagons are furnished by Sam Smiser, and the mules come from Red's Meadows Pack Station.

Left: **Bobby Tanner** *driving the 20 mule team while riding the near wheeler.*

The **20 mule team** *in the Saturday parade. The 80 foot-long chain running between the mules is clearly evident in this picture.*

Bobby Tanner *demonstrates a left turn with the 20 mule team. In order to keep the wagon from cutting across the corner, the three pairs of mules in front of the wheelers (mid-teams) step over the center chain and pull the wagon at almost a right angle to the turn. When the wagon gets up to the corner the mid-teams step back over the chain and continue to pull normally. These mid-team mules perform this maneuver without missing a step.*

HALTER CLASSES

There are a full compliment of halter classes at Bishop, with the mule classes divided by age and by type of mule (quarter type, thoroughbred type, colored, and draft) and the donkey classes divided by age and size. But, in contrast to the Columbia Mule Day halter show, there were no entries in the 1993 Draft Mule Halter Classes at Bishop.

***Aristocrat's Levi**, owned and shown by **Marr Olsen**, of Petaluma, California, was named 1993 World Champion Jack. He placed 2nd in his class, Mammoth Jacks 54" and over, 3 years old and older.*

***Wendy Dexter** of Winnemucca, Nevada, took first place in the aged Mammoth Jack Class with **Cactus Jack**, owned by Robert Dexter. I didn't understand how Cactus Jack was beaten by Aristocrat's Levi in the Championship Class, after they were placed the other way in their Jack Class. This must have been a judging error, as Jack did not appear lame or otherwise different when he came back to be judged for Champion.*

Gypsy Toi*, a 3-yr-old mare mule owned and shown by* **Renee Delpino** *of Riverside, California, was the winner in five separate halter classes! Gypsy Toi is a red dun molly by the well-known jack, Big John, out of a world class AQHA mare mare named Tameo. Shown at halter since her yearling year, she has been beaten in her class only by Poco Tasha (which we will see on page 82 in the Shelbyville story).*

Winner of the 2 year old quarter horse mule type class was ***Norma Jean Reily****, owned and shown by* **Dorothy Bayer** *of Antioch, California. Norma Jean was a 20th wedding anniversary present as a yearling, and is by Valentino, a red roan Mammoth Jack out of a palomino Quarterhorse mare. This was Norma Jean's first time showing against mules, after previously doing very well against horses in local horse shows.*

The 1993 World Champion Saddle Halter Mule, **Maynard's Miss Sheba**, *a 5-yr-old mare owned and shown by* **Tonya Albertson**. *Sheba's halter show record is almost unbelievable for such a young mule, with two World Championships, two National Championships, and fifty-five Grand Championships.*

Tonya trains mules of her own and for others at her Pine Meadows Mules Ranch in Winnemucca, Nevada. She also maintains a mule breeding program using Dry Gulch Maynard, Southsides Hot Shot, and Black Bart Mammoth Jack bloodlines on Quarterhorse mares with proven halter and performance abilities.

Tonya *and* **Miss Sheba** *in the Western Riding-Green Class. 1993 was Sheba's first year under saddle, and she started off in a big way with four Championships and two Reserve Championships in the Green Western Division.*

The 1993 Reserve World Champion Saddle Halter Mule, ***Sweet Water LaVerne****, owned and shown by* ***Judy Rose*** *of Escalon, California. At Laverne's head is* ***Larry Gimple*** *from Chino, California, judge for the Western and saddle mule halter classes. LaVerne was a birthday present to Judy as a 2-yr-old in 1978, from her husband, Gary. She won her first championship as a 3-yr-old, and hasn't stopped yet at age seventeen. LaVerne is out of a Three Bars Quarter mare and by a mammoth jack owned by George Chamberlain.*

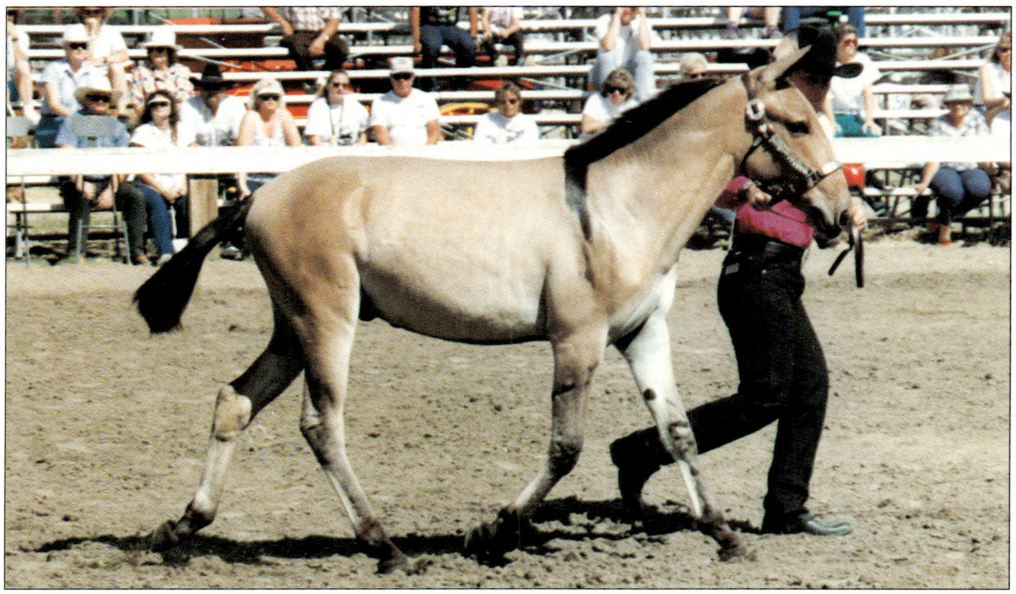

Kelly Baker *showing her yearling buckskin mule,* ***Actions Buckwheat****. Kelly bred Buckwheat at her ranch in Chino, California. He is by Action Jackson, her 14.1 hand red-gray roan jack and out of a registered quarterhorse buckskin mare. Kelly sold Buckwheat to Bob & Diann Walker of Mansfield, Texas, during the Bishop show, and now is looking forward to showing his full brother at Bishop in 1994.*

The Yearling Jack Class was won by ***Eagle Canyon Jack Benny****, owned and shown by* **Glen West** *of Las Vegas, Nevada.*

The 3 yr-old donkey class was won by ***Danke Schon - Bandit****, owned and shown by* **Valerie Percival** *of Sebastovol, California.*

Mike Trueblood *showed his white yearling mule,* ***Spuds Mackenzie****, in the Quarter Horse Type Class. Mike is from Lincoln, California.*

ENTERTAINMENT ACTS

The sponsors of Bishop Mule Days do a very good job of entertaining the grandstand crowd, both with the classes and with other acts which help to fill in the time when jumps have to be put up or taken down, obstacles have to be removed, or there are other gaps in the program. The clowns are excellent, and the dual announcing team of Bob Tallman and Bob Feist are first class. All in all, it is a very professional operation.

Tom Feller *retired from 17 years as a professional rodeo clown and is now Event Marketing Director for Tony Lama Boots. But once each year, during Bishop Mule Days, he again becomes a clown. Here he is riding his trusty buffalo, "Chip", an outfit that strikes fear into most of the mules that Tom confronts in the ring.*

During the Saturday and Sunday shows the crowds were thrilled by the trick riding of **Tad Griffith**, *a world famous stunt rider. Here he is doing some Roman riding on a team of Percherons.*

Another clown, **Donnie Landis**, *demonstrates how to milk a jack donkey by pumping the handle (tail) up and down. I was skeptical, but by golly, it worked.*

CATTLE WORKING CLASSES

The Bishop show has a full compliment of cattle working classes — cutting, roping, and penning. These classes are fun and exciting, and a real challenge to both exhibitors and photographers.

Walter Nunn, *of Bryan Texas, riding* ***Clementine****, won the title of World Champion Cow Working Mule and Rider. You can almost feel the intensity of both Walter and his mule as they approach a group of calves, deciding which one to cut out of the group. It is important to pick a calf that will try hard to return so that your mule will give a good performance. Behind Walter is* **Daniel Martin** *on* ***Matilda****, holding the cow group together. At the left we see Walter and Clementine keeping the calf they selected away from the rest of the group.*

Tucker Slender, *from Arcadia, California riding* **Judy**, *in the Steer Stopping Class.*

Albert Andrews, *of Laton, California, riding* **Whiskey**, *took second in the Cow Cutting class.*

The **Martin Brothers** *presented a handsome picture as they rode out to meet their foe in the Cattle Penning Class, but things soon fell apart as the cattle scattered and would not cooperate.*

There were many lady exhibitors in the cattle classes. Here are two views of **Crystal Elzer** *cutting cows with* ***Teaspoon****. For a shock turn back to page 32 to see this same pair in the side saddle class. These two will ride in the 1994 Tournament of Roses Parade — Crystal's fifth Rose parade and Teaspoon's first.*

This is **Jo De Collins** *of Roseville, California, riding* ***Color Me Grand****. They took third in the Cow Cutting Class.*

Eric Midtsjo *of Norco, California, didn't compete in the cow cutting, but he sure looked like he could. Instead he helped to herd and contain the cows while the others did their cutting.*

Two shots of **Ann Wood** riding ***Easy Ed's Rita*** *in the Cow Working — Green Class. Rita is owned by* **Kick Ass Mules** *of Las Cruces, New Mexico. This pair took first in the English Working Hunter — Green Class on their way to being Reserve World Champion Green Mule — English.*

Right: **Chuck Reid** *of the Burnt Leather Ranch in McLeod, Montana, on* ***Southside****, a 7 yr-old molly mule by Dry Gulch Maynard out of a registered racing quarter mare. Southside won the title of Reserve World Champion Western Performance Mule at the 1993 Bishop show.*

Walter Nunn *on* ***Scott's Sagebrush Molly****, owned by* **Lisa Fergason** *of Sangor, Texas. This pair took first in the Team Roping Class. Molly is a 13 yr-old, 15.3 hand mare by the mammoth jack Scott's Sagebrush and out of a registered thoroughbred mare.*

WESTERN RIDING CLASSES

The Bishop Western Classes include Western Pleasure, Reining, Trail, and Western Riding. For the purpose of this section we will also include the Gymkhana classes (Barrel Racing, Polebending, and Keyhole Race) and the donkey riding classes

Herb Kunz *of Goleta, California, riding* ***K-Mules Dandy*** *in the Trails Class.*

Two photos of **Megan Greslie** *riding* ***Sarah Burnett*** *to a 2nd place finish in the Youth Trail Class, 13 years and under. Sarah Burnett won the Open Trail Class (36 entries) with Megan's sister,* **Miki Greslie** *riding. The Greslie's are from Sun Valley, California. Sarah Burnett is an 11-yr-old mare by Scott's Jack out of a quarter mare. She was the AMA High-point Youth, Champion Performance, and Overall Highpoint Mule in 1988, and has won many other Championships in the past five years.*

Loyd Hawley *of Prairie Grove, Arkansas, riding* **Clemantine My Darlin**, *won the Polebending, Barrel Race, and Keyhole Race Classes, making it a clean sweep to become 1993 World Champion Gymkhana Mule. This pair won the same honor at Bishop in 1991, 1993, 1985, and 1987, and won Reserve World Champion in 1989 and 1992. Clemantine is a 10-yr-old mare out of a running quarterhorse dam.*

There were three entries in the Donkey Pleasure Class. **Meredith Hodges** *and* ***Little Jack Horner*** *were the winners. Here they are walking to the ribbon presenter to get their blue ribbon.*

Second place went to ***Rooster****, ridden by* **Charles Moore**.

In third, but with a very nice ride, was **Fran Howe** *of Bennett, Colorado, riding* ***Blue Zebulon****.*

Chuck Reid *riding* ***Southside*** *in the Western Riding Class. There were 23 entries in the Western Riding Class.*

Kathy Gardinier *and* ***Easy Auge*** *go through the gate in the Western Riding Class. Easy Auge is owned by Kick Ass Mules of Las Cruces, New Mexico, and took first in the 4 yr-old halter class.*

Bobbe Johnson *on* ***Rose Marie*** *in the Western Pleasure Class. There were 40 entries in this class, plus another 25 entries in the Green Western Pleasure Class. Bobbe is just one of many exhibitors who rode her mule both English and Western. She is pictured in the Dressage Class on page 32.*

The Youth Pleasure Class, 13 and Under, was won by **Adrienne Shor** riding ***Apples Peaches and Cream***.

Katie Jones *of Palmdale, California, rode* ***Ginger Luck*** *to a 3rd place ribbon in the Youth Pleasure Class, 13 and Under.*

The Keyhole Race involves entering an area marked off like a keyhole, spinning around, and racing for the finish line without going outside the chalk marks. Here **Manuel Garcia** *of Hanford, California, and* ***Tawnya*** *are making a fast exit from the keyhole area. They took second place in this class.*

MULE PULLING

On Sunday morning they have a mule pulling match. There are four weight catagories, from Mini-mules (team weight of 1001# or less) to Heavy mules (team weight of 2301# and over). Pulling must not be a real big sport out west, as there were a total of only 9 teams entered for all four classes.

Jon Martin *driving his small pulling team,* ***Maude and Jack****, to a win in their class. The Martin Brothers, from Nampa, Idaho, brought 19 mules and 27 family members to the 1993 Bishop Mule Days. There was a Martin Brothers entry in most classes, except for the English riding classes.*

The sleds looked empty, but those little blocks were really heavy. They loaded these blocks on and off the sleds with a hydraulic crane.

Larry Creason *drove the Vivian McCall (Oro Grande, California) heavyweight team of* ***Sid and Mike****.*

MULE PACKING CLASSES

The packers are boss — Number 1 — at Bishop. It started that way in 1970 when the first Bisop Mule Days were set up to showcase the mules which worked at the pack stations in the nearby Sierra Mountains, and it remains that way some 23 years later. Some of the other mule factions have crowded the packers some, and will continue to do so in the future, but the packers will probably remain "King" at Bishop for many years to come.

There are 15 independent Pack stations in the nearby mountains, plus many pack strings owned by the Federal Government for use in the National Forests. Many of these outfits, plus some from out-of-state, come to show their packing skills and the training of their mules at Bishop. **There are 19 separate classes for the packers**, many of which are held during the "prime time" evening shows in front of the grandstand. It's hard to believe there are 19 different ways you can design a mule packing class — but they have done it at Bishop.

Scott Jones *won the title World Champion Packer at the 1993 Mule Days. Here he is competing in the Individual Scramble, a class in which he took the blue ribbon. Scott, in the purple shirt, must tie an odd mixture of pipes, barrels, and other objects on the back of his two pack mules, mount his horse, and lead the loaded pack mules across a finish line. The first packer to complete this task, without his load coming loose or undone, is the winner. Above and to the right we see Scott completing his load on the first pack mule, with the second mule waiting patiently to be loaded. Clay Rose is holding Scott's horse, and the mule he is packing.*

Oops. In the frantic haste to complete the packing Scott has lost his hat.

Scott Jones leads his two loaded pack mules to the finish line.

Jon Martin *leads the Martin Brothers pack string out of the showring in the Team Packing Scramble class. This is similar to the individual Scramble Class, except that there are more people and more pack mules involved.*

Two shots of **Jon Martin** *leading the Martin Brothers pack string through the obstacle course in the Pack Train Class. The Martin Brothers have an unusually attractive string of snow-white pack mules.*

At the left we see the **Agnew Meadow's Women** *team, one of the two women's pack teams in the show, going through the serpentine obstacle course.*

Here is the same pack team coming back into the arena for another class. No, that isn't a zebra — only the same mule you see above but with a new paint job.

On Saturday night **Todd Howell** *from the* **Fred Harvey Livery** *outfit took his regular weekly bath on muleback, in front of the grandstand, during the Team Packing Comedy Class. The fellow on the left, riding a mule, is* **Casey Murph**, *who makes his living packing supplies down to Phantom Ranch at the bottom of the Grand Canyon.*

The **Rock Creek Women's** *pack string did a very nice job going through the obstacles despite the presence of the clown, Chip, outside the barrels.*

Announcer **Bob Feist** *did much of his announcing while riding* **Samantha Harlan's** *mule,* ***Sissy T****. Bob thought the Rock Creek Women's pack string worked so well that he would try it. Here he is starting out, leading the Rock Creek string through the gate.*

Sissy T*, however, was scared to death of* **Chip***, and she spooked whenever she saw him. Bob's nice demonstration ride fell apart soon after this picture was taken, when Sissy T noticed Chip and took off in the other direction. The crowd loved it.*

MULE AUCTION

There is a mule auction on Friday afternoon. The flyer for the 1993 auction listed 26 animlas, a rather modest offering compared to the mule auctions held in Tennessee and Missouri. A good number of the animals offered were passed out of the ring without being sold.

This strawberry roan 3 yr-old molly mule from **R and B Saddle Mules** *brought $3,500, but was passed out without being sold as the owner needed $4,000.*

Dick Reese *brought this 4-yr-old Tennessee Walker molly mule from Tennessee, but it too was passed out without being sold after reaching a bid of just $1,250. These gaited mules are not real popular in California, but are quite popular in Tennessee. See page 78 for some pictures of the Walking mules at the Shelbyville Show.*

BISHOP MULE DAYS PARADE

The 1993 Mule Days Parade, with 151 pre-registered entries, started out at 10:00 am on a beautiful, sunny day. Most of these entries were in place an hour earlier, for judging.

In contrast to the Columbia parade which included many horses, the only horses I noticed in the Bishop parade were four Percherons hooked to a wagon and some saddle horses leading mule pack strings. There was one zebra entry and a few llamas — all the other entries were all mules or donkeys. There are no motorized vehicles in the Bishop parade.

The parade route is fairly long, and the crowds are not excessive, which makes it relatively easy to find a good spot to view the parade. Just bring a lawn chair, wear a hat, and enjoy.

Lee Roeser's *pack string of nine black mules, each with two American flags flying on top of their white packs, made an impressive and lovely sight whenever they appeared, such as the beginning of the Mule Days Parade and during the opening ceremonies of the evening shows in front of the grandstand.*
Lee and his wife, Jennifer, operate the **McGee Creek Pack Station** *near Mammoth Lakes, California, where they offer personalized wilderness vacations for families, individuals, and groups in the nearby mountains.*

Ben York *of Alpine, California, rode his horse, Leverage, in the parade leading his two pack mules,* **Texaco** *and* **Molly***. Dr. York, a retired veterinarian, and these three animals rode the entire length of the Pacific Crest Trail, some 2,600 miles, in the summer of 1992. They started on the Mexican border April 21st, and ended in Manning Park, Canada, on September 21st, a truly remarkable feat. As president of the Back Country Horsemen he wanted to see for himself the condition of the trail. He found what he suspected — incredible beauty but serious neglect of the trail itself — and hoped that his participation in the Mule Days activities would call attention to this neglect. And it did!*

Cindy Powell *of Hunter Mountain Farm from West Augusta, Virginia, presented a pretty picture riding* **Banjo***, a 9-yr-old red dun horse mule.*

Dr. York enjoys the sunshine as he waits for the parade to begin.

This is the Lane Ranch, Lancaster, California, hitch of four mules pulling a handsome stage coach — **George Lane** *is driving. The movie industry often uses Lane Ranch vehicles and animals in their westerns.*

Herb Kunze *of Goleta, California, cut a fancy figure on his saddle mule.*

Heidi Sylvester, *dressed in a lovely costume, was waiting for the parade to start. The mules,* **Ashley** *and* **Scarlet**, *are owned by Heidi's gramdfather, Rod Stillion of Palmdale, California.*

There were over a dozen pack strings in the parade. The top two strings on this page belong to the **Yosemite National Park**. *(Yosemite is quite close to Bishop. We tried to go there after the Bishop Show but the only road from the west, where we were, was still closed due to heavy snow the early part of June.) The string at the bottom is the* **Rock Creek Women's Team**. *The middle and the bottom pictures illustrate actual loads which would be carried by a pack train, whereas the top string is dressed for the parade.*

Ray Muller *of French Camp, California, did demonstration drives with his ten-up of mules in front of the grandstand as well as driving them in the parade.*

Samantha Harlan *of Marysville, California, riding* **Sissy T***, won a first prize in the parade competition. Sissy T is the mule which Bob Feist borrows to announce from during the grandstand shows as pictured on page 63.*

Reg Coldren *of Goleta, California, dressed in a Mexican costume, also won a first prize in the parade competition.*

Von Twitchell *of Marysville, California, cut a fancy figure on his mule in western tack.*

Crystal Elzer *of Scotts Valley, California, rode side-saddle on her mule* ***Teaspoon****.*

There were a few llamas in the parade. They are a photographer's dream. Their ears always seem to be in the alert position.

Will Willeford *of Yucca Valley, California, rode in the parade as a Fancy Western Man.*

Diane McElvain *from Sun City, Arizona, on her Appaloosa mule.*

The final entry in the parade, number 151, was **Tonya Albertson** *of Winnemucca, Nevada, riding* ***Maynard's Miss Sheba*** *while wearing an indian maiden parade outfit.*

The Smizer Freight Service *had two entries in the parade, both with unusual vehicles. This four-up of draft mules is pulling an antique beer delivery wagon.*

Freddie Valentino *of Lancaster, California, drove a six-up in the parade. All of these mules were bred, raised, and trained by Freddie. The wagon is an old breaking wagon, and the harness is a replica of the type used on the 20 mule teams 100 years ago.*

Gerri Goers *rode her Appaloosa horse leading her Mammoth Jack,* ***Ima Too Too Handsome Jack****.*

This entry, from Jamul, California, was called Mail Order Brides.

Maynard's Miss Sheba*, the 1993 World Champion Saddle Halter Mule.*

Aristocrat's Levi*, 1993 World Champion Jack, feels good about being part of Bishop Mule Days.*

Two views of the Calsonic Arena, the indoor arena where most of the Great Celebration Mule Show classes are held. This is a new facility (1989) of world class stature. The ring area is 300′ x 150′, surrounded with plenty of comfortable seating.

The Great Celebration Mule Show

Shelbyville, Tennessee

Shelbyville is in western Tennessee, about 50 miles south of Nashville and about the same distance east of Columbia. It is in the heart of the Tennessee Walker country, and the Celebration Grounds in Shelbyville is the home of the annual National Tennessee Walking Horse Celebration for the ten days prior to the Labor Day holiday.

The horse show facilities at Shelbyville are excellent, and there is a staff of professional horse show people year-round who schedule other events, in addition to Tennessee Walker shows, to better utilize the facilities. In 1992 they started The Great Celebration Mule Show.

The folks at The Celebration offices are frank to admit that they knew, and still know, very little about putting on a mule show. They have enlisted the advice and help of mule people to establish and operate this show. A local influence can be seen in a strong schedule of draft mule classes. A California (and saddle mule) influence has been provided by Judy Rose, a noted mule breeder from Escalon, California, who has advised the Celebration people and judged the saddle mule classes for both 1992 and 1993.

The Shelbyville show is more like a major conventional horse show than is Columbia and Bishop — most classes are held indoors and there is no parade. The Calsonic Arena in which the classes are held is a world class facility, and the outdoor practice rings and the stabling is more than adequate. The personnel running the show are professionals, and the auxiliary facilities (speaker systems, food facilites, vendor areas, parking, etc.) are excellent.

Without a long tradition to hamper them, the Shelbyville show could well become the modern mule showcase of the future. The desire and facilities are there. All that is needed is increased education of the eastern mule people on what is expected at a first class show and some participation from the western mule people.

Shelbyville could (should?) become the permanent home for the National American Donkey & Mule Society Show. Which brings us back around to the problem of having two mule organizations, and two "National" shows — an issue on which which I am not qualified to comment except to say that it is unfortunate.

Ricky Davis *from Carrabelle, Florida, riding his mule,* ***Pine Log Red****, in the practice ring next to the Calsonic Arena. Pine Log Red earned* ***High Point Mule*** *honors at the 1993 Great Celebration Mule Show.*

Lisa Ferguson *from Sangor, Texas, with* **Scott's Sagebrush Molly**. *They won the Senior Western Pleasure Class, and are shown here being presented with their ribbon by the ringman,* **Buddy Black** *(who is driving his team at Columbia on page 11). Lisa and Molly also took first in the Senior English Pleasure Class and second in the Reining Class. We also saw Molly in the Bishop show on page 53, helping Walter Nunn take a first in the Team Roping Class.*

Ed Huie *from Ila, Georgia, rode* **Eunice** *to a first place in the Junior Western Pleasure (4 and under) Class. Eunice also was first in the Saddle Mule to Halter, Four & Over Class.*

This is **Kelly Williams** *riding her mule,* ***Jungle Jane***, *in the Junior Western Pleasure Class. Kelly works as an Equine Marketing Specialist for the Tennessee Department of Agriculture, and was very involved in the Celbration Show. Jane is a 4-yr-old mare by a mammoth jack and out of a Tennessee Walking Horse mare.*

This is **Christine Duval** *of Onalaska, Wisconsin, first practicing outside and then winning the Hunter Jumper Class with her 11 yr-old mule,* ***Athena S. Vasquez****. Athena started her career in roping and cow events. Athena came to Wisconsin four years ago and was switched over to English events. Athena was recently awarded the ADMS Versatility Hall of Fame Award, with most of their points accumulated in open horse shows against horses in Minnesota and Wisconsin. At Shelbyville they won the Hunter Jumper Class, took 2nd in English Pleasure, and 3rd in Western Pleasure.*

Ricky Davis *from Carrabelle, Florida, rides* ***Pine Log Red*** *across an obstacle in the Trail Class. Red was the high point mule at the 1993 Celebration, winning the Reining Class and placing high in many other classes, both English and Western. Ricky bought Red unbroken at the Dickenson (Tennessee) Sale in January, 1993, and was shown at Shelbyville with just 6 months training.*

Calleton Evans *and* ***Dolly*** *in the Western Pleasure Class. Dolly is an 8-yr-old mule with four white stockings — very flashy — with a easy way of going and a sweet disposition. Calleton, age 64, enjoys trail riding with Dolly in the Shawnee National Forest.*

The Tennessee Walking Mule Classes were exciting and beautiful to watch. These mules are flat-shod, and move with a smooth, easy gait. At the left we see **Eddie McCrary** *of Lebanon, Tennessee, practicing outside with* **Little Darling**, *a mule owned by* **Hub Reese, Jr**. *Later that evening Hub rode Little Darling in the 5 Year and Older Tennessee Walking Mule Class, and took first place. At the right we see Eddie riding* **Rambling Rose** *to a win in the Four & Under Tennessee Walking Mule Class. Rose fairly flew around the ring, cheered on by the large crowd.*

Rufus Reese *of Gallatin, Tennessee, rode* **Hoppin Fred** *to a first place finish in the Tennessee Walking Trail Pleasure (Western Tack) Class. Here he is leaving the ring with his ribbon and trophy.*

Bill Moore *raises Tennessee Walking Mules on his Stepping Out Farm in Shelbyville, Tennessee. Here he is riding a black mule called Boot Scootin' Boogie in the Five Year and Older Tennessee Walking Mule Class.*

Wynne DiGrassie *from Staunton, Virginia, on* **Crackerjack Surprise**, *a 15 hand, 4-yr-old mare mule out of a registered Palamino Quarterhorse. Crackerjack was third in the Aged Saddle Mule to Halter Class (26 entries).*

This is **Vicki Upton** *from Springerton, Illinois, with her weanling filly called* **Two Bits Crickett**. *The Uptons got hooked on saddle mules four years ago, and are now breeding their own. Crickett is by Eagle Rest Two Bits King Pen and out of a quarter mare Suzzette, herself a daughter of AWHA champion Sun Up Jack. After the Celebration Vicki took Crickett to the Indiana State Fair where she was Reserve Chanpion Saddle Mule.*

Joe Gillespie, Jr., *from Brookville, Indiana, riding his 5 yr-old mare,* **Matilda Bay**. *This pair placed fifth in Western Pleasure (22 entries). Matilda is also out of a quarter horse mare.*

This is **Lucky Three Stardust**, *a 9-yr-old mare out of a quarter horse mare, being ridden by* **Wynne DiGrassie**. *Stardust is owned by* **Hunter Mountain Farm** *in West Augusta, Virginia. Stardust was the Celebration High Point Mule in 1992. In 1993 she was second in halter, third in English Pleasure, and fifth in Jumping.*

Cindy Powell *of Hunter Mountain Farm in West Augusta, Virginia, driving* **Banjo***, a 9-yr-old red dun mule out of a quarter horse mare. Cindy and Banjo won both the Single Horse Driving Classe.* **Wynne Digrassie** *took a first with Banjo into the Side Saddle Class.*

Rilla Reese *drove* ***Jargo*** *to a win in the Youth Driving Class. Riding with Rilla is her father,* **Rufus Reese** *of Gallatin, Tennessee. Rufus and his borther, managed a Mule Sale on the Celebration grounds on Saturday morning. Jargo, owned by the Reese Brothers Mule Company, was a dominant mule at the Celebration, entering many different types of classes, often placing in the top three.*

This pair of fancy mules came all the way from Hickory, North Carolina, to be seen on the Celebration grounds and to compete in the Chariot Race Class, where they placed second. They are 2 and 3 yr-old horse mules owned by **Worth Whisnant***, the man who was driving a mule flanked by two Brahaman steers at Columbia pictured on page 24.*

Jenny Brooks *drives her 9-yr-old molly mules,* ***Kate and Nell****, around the grounds in preparation for the driving classes. Riding alongside Jenny is her father,* **Howard Brooks, Jr.** *of Chattanooga, Tennessee. Jenny also drove Kate in the Single Driving Class where she placed second.*

Wayne Hussey *with a pair of 5-yr-old molly mules,* ***Pat and Lou****, owned by* **James Van Camp** *of Carthage, North Carolina. This outfit took four firsts, one second, and one third at the 1992 National Mule Show in Columbia, South Carolina. After the 1993 Celebration Show they went on to the prestigious driving show in Devon, Pennsylvania, where they placed second in the Commercial/Farm Class. At Devon they were noticed by others and invited to perform exhibition drives at the World Pair Championships in Gladstone, New Jersey, in October, 1993. This turnout is doing a fine job in spreading the word that mules can hold their own in the equine driving community.*

Amy McLean *of Bostwick, Georgia, was a very busy girl during the Celebration. At the left she is practicing the jumps with* **Secret of Victoria***, a 5-yr-old mule which Amy has owned for 3 years, and has trained herself (with a little help from Dad). Amy and Victoria participated in many of the English and Western Classes, and took first in the Junior English Pleasure Class. Amy also rode Fido, a Tennessee Walking Mule, to a second in the Ladies Side Saddle Class. Above we see Amy doing a little practice driving with* **Polly***, and her Dad, Ray McLean, riding alongside.*

Loyd Hawley *of Prairie Grove, Arkansas, won Champion Saddle Mule at Halter with* **Poco Tasha***, a 3-yr-old molly mule by a small Mammoth jack and out of a quarterhorse mare. Loyd also won the Pole Bending and Barrels Classes riding a black mule called Dipstick.*

The draft hitch show was similar to the show we saw earlier at Columbia, except that it was 50 degrees warmer and the mules were shed out — but the exhibitors and mules were largely the same. Here we see the team class lined up before the exhibitors take their individual workouts. There are 11 teams in this picture.

Thomas Summers *of Morrison, Tennessee, again dominated the draft hitch classes, winning all five classes he entered plus both log skidding classes. Above we see* **Bob Cantrell** *practicing with one of the Summers' teams,* **Bell** *and* **Bird**. *With Bob is his son, Jeff. The mule at the right is* **Kate**, *one of the Belgian mules in Tom's hitch.*

Greg Tabor *swings his team to the left in the practice ring. This team,* **Kate** *and* **Ida**, *placed second in the Team Class. Greg is from Scottsville, Kentucky, and with him on the seat is his wife, Teresa.*

Judy Rose, *from Escalon, California, was the driving force for the saddle mule classes at the Celebration. She judged the saddle mule classes in both 1992 and 1993, and helped the Celebration folks set up the class schedule. Judy does an excellent job of explaining the rules, and what she expects, to the exhibitors.*

Jargo, *a grey molly mule owned by the Reese Brothers Mule Company, did almost everything at the Celebration — English Riding, Western Riding, and Driving. Here she gets a wash job by her most frequent rider,* **Sarah Flynn** *before going into the Youth Showmanship Class, which Sarah won. The circle photo shows Sarah before she entered the side saddle class where she placed fourth.*

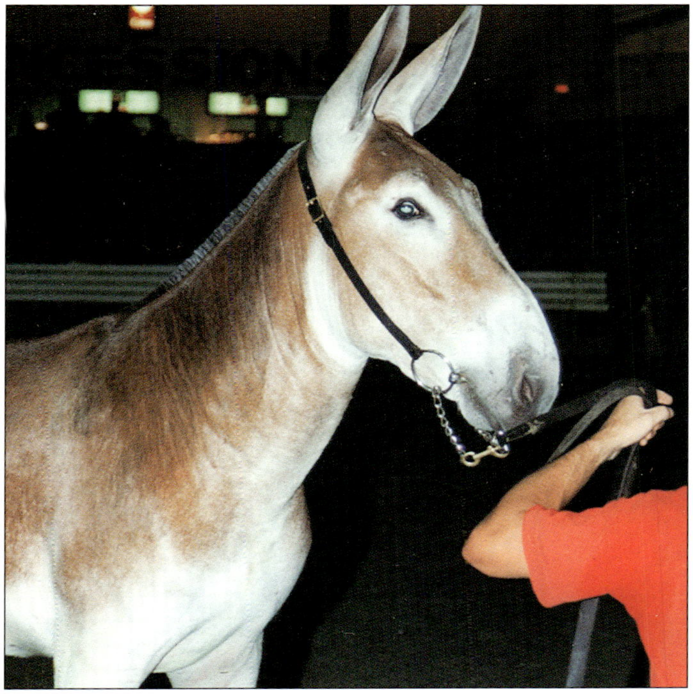

The Champion Jack was **Charlie**, *a 2-yr-old owned by* **Bill and Will Trice** *of College Grove, Tennessee. The Trice's pretty well dominated the Donkey halter classes, winning every Jack class and either winning or placing second in every Jenny class.*

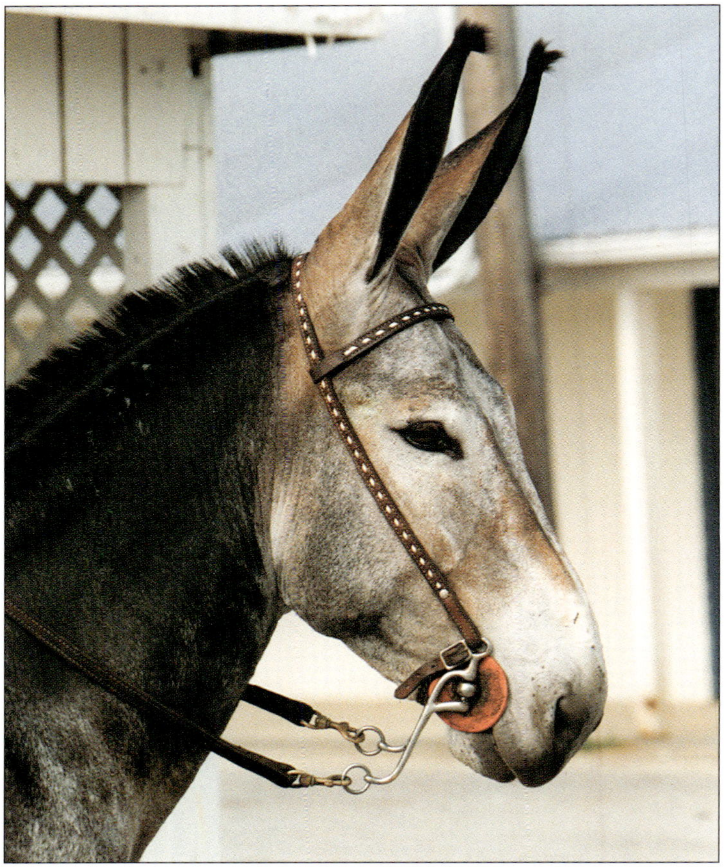

This is the aged Jack, **Old River**, *owned, shown, and ridden by* **Stanley Nix** *of Lewisburg, Tennessee.*

Will Trice *is showing* **Kate**, *the winner in the 3-yr-old Jenny Class.*

Grand Champion Draft Mule at Halter was the aged mare, ***Lize****, owned by* **C. C. Sewell** *and* **Hub Reese**. *In the left-hand picture we see* **Bob Kaiser***, Lize, Hub Reese, and C.C. Sewell. Bob is Mr. Sewell's grandson, and he handled Lize in her class. Lize was also Grand Champion at the Columbia show (see page 14), but now she has a little less hair and more shine than she had in April.*

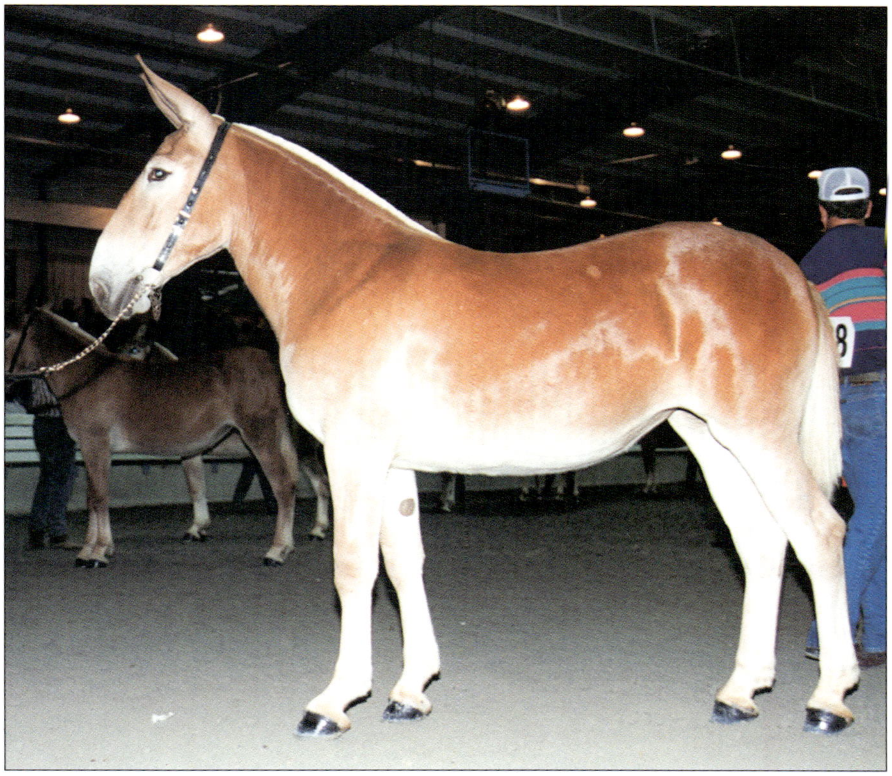

Two of the lovely molly mules from the yearling class. On the left is ***Judy****, the second place entry from* **L.C. and Greg Tabor**. *The head shot on the right is of* ***Jane****, the first place entry from* **Clarlie Reed** *of College Grove, Tennessee.*

Saturday morning is devoted to a Mule Auction put on by the **Reese Brothers Mule Company**. *This fancy pair of 7 yr-old draft mules brought a bid of $2,400 each, but it wasn't enough to buy them. There is a strong demand for good draft mules.*

The winner of the Champion Mare Mule Pair at Halter Class was ***Pat and Jane****, two aged molly mules owned by the* **Reese Brothers Mule Company**.

The Bishop and Celebration shows provide an excellent opportunity for breeders, trainers, and owners to showcase their good saddle mules. Although the market remains strong for draft mules the potential for growth is much greater for saddle mules. It is, in fact, a seller's market limited only by the number of good saddle mules that are being bred.

Dr. Joseph Hunt *of Carleton, Michigan, is a tireless promoter of the mule and a faithful exhibitor at the Michigan Great Lakes Show. Here he is exercising his four before entering the Coliseum. Note the differences in the types of harness used by these two exhibitors on this page.*

Jack Strode *of McMinnville, Tennessee, leaves the Coliseum after winning the Four Mule Hitch Class. (This picture, and a few others in this chapter on the Michigan Show, were taken in 1991. All the other pictures in this book were taken in 1993.)*

Michigan Great Lakes International

Each October, on the grounds of the Michigan State Fairgrounds in the city of Detroit, there is a draft horse show called the Michigan Great Lakes International. It is the largest draft horse show in North America, with over 1,000 Belgians, Percherons, Clydesdales, and Shires competing for $60,000 in premium money. There are usually over 100 exhibitors in both the Belgian and Percheron classes, and the show extends over a period of four (very full) days.

Five years ago the International show management added classes for mules. Inasmuch as this is a draft horse show, the mule classes are for *draft* mules. The mule classes are not really filled, with only about six exhibitors in attendance each year. This is a good opportunity for draft mule breeders to showcase their animals.

This is **Vernon Cornett** *of Jeffrey, West Virginia, with his six of very large draft mules.*

Don Bales *of Sabula, Iowa, drives his three abreast team at the Detroit International. The hightlight of the show for Don was when he won the Six Horse/Mule Driving Competition with a 3 minute free drive, including several figure eights at a gallop, beating eight other exhibitors.*

Below: The lead mules on **Dr. Hunt's** *team.*

Appendix

For additional information on the shows pictured in this book you can contact the show management at:

Columbia Mule Day
P. O. Box 66
Columbia, TN 38402
(615) 381-9557

Bishop Mule Days
690 N. Main St.
Bishop, CA 93514
(619) 873-8405

The Great Celebration Mule Show
The Celebration
P. O. Box 1010
Shelbyville, TN 37160
(615) 684-5915

Michigan Great Lakes International
c/o Dan Creyts
7775 Davis Highway
Charlotte, MI 48813
(517) 322-0315

Bob & Mary Mischka publish and sell books and calendars which feature mules. For a copy of their catalog you may contact them at:

Mischka Farm
N8246 Esterly Road
Whitewater, WI 53190
(414) 473-5595

Index of Exhibitors